For my children and
all other children of the world.

In a big, busy city in the USA

there lived a little boy who wanted to learn to count one day!

Thinking of things to use and help him count, the little boy decided on fish that swim about.

Once certain and for sure, the little boy decided to travel to the local pet store.

Once arriving at the pet store, the little boy saw cages for cats and dogs, birdseed and more!

1

Excited, the little boy focused on his mission, quickly walking over to the fish section to count ten different specimen.The first was the most amazing, one Blue Tang fish.

Second, on the list were two Rainbow fish.

2

**Third, he did not dare miss,
three Marine Angel fish.**

3

Fourth, on the list were four Yellow Tang fish.

4

5

Fifth, it only made sense to count the five Gold fish.

Sixth, very brightly colored and decorative were six Mandarin fish.

Seventh on the list, the littleboy went on to count seven Yellowfin Surgeon fish.

7

Eighth, with heads red and enormous, the littleboy was quick to notice eight Flowerhorn fish.

8

9

Nineth, truly one of nature's gifts, the littleboy counted nine special Parrot fish.

10

Finally, was the tenth in a small collective, ten Copperbanded Butterfly fish.

Thrilled about the fun he had that day, the little boy learned to count to ten in a very exotic way!

The End

ISBN 978-0-578-46604-0

The text of this book is set in Verdana Bold.
The illustrations are in pencil, marker, and watercolor.

Printed in the USA

www.ingramcontent.com/pod-product-compliance
Lightning Source LLC
Chambersburg PA
CBHW061139030426
42334CB00004B/94